Starry Night™ Observation Exercises

William J. F. Wilson

T. Alan Clark

W. H. Freeman and Company
New York
www.whfreeman.com/astronomy

Supplements Director: Patrick Shriner
Associate Editor: Matthew P. Fitzpatrick
Project Editor: Jodi Isman
Administrative Assistant: Ceserina Pugliese
Composition and Design: Christopher Wieczerzak

ISBN 0-7167-3516-4

Printed in the United States of America

First Printing, 1998

Observation Exercises

Preface

By special arrangement with Sienna Software, the CD-ROM accompanying each copy of *Universe,* Fifth Edition, by William J. Kaufmann (late) and Roger A. Freedman, features a student version of the acclaimed software, Starry Night™. The combination of the best-selling introductory astronomy text (long praised as exceptionally authoritative, captivating, and vivid) and this astonishing virtual planetarium makes this an unbeatable learning package.

Some of the most striking examples of the powerful *Universe*/Starry Night™ partnership are the observation projects in the text, which students can perform using the software on the CD-ROM. This exercise booklet features more comprehensive observation projects that are written specifically for Starry Night™, thus providing many more opportunities to explore the phenomena of the cosmos through the window of the computer.

Starry Night™ Deluxe is an enhanced version of Starry Night™ that lets you go anywhere in the solar system, features over 19,000 stars, adds the ability to create your own customized planets, has new options, and much more.

Starry Night™ Deluxe is available as an upgrade to users of the student version for $59.95. To upgrade, just call or contact Sienna Software through the web and ask for the "W. H. Freeman upgrade."

800-252-5417 (toll free)
416-410-0259 (international), or
www.siennasoft.com/whfupgrade

1

Stars and Constellations: The Big Dipper—Guide to the Sky

Stars are scattered across the nighttime sky in a way that can be bewildering to a person who is unfamiliar with the patterns that are formed. However, the sky is like any unfamiliar location—once you are able to recognize a few prominent landmarks, you can begin to find your way around with relative ease, and filling in the areas between the landmarks can be quite enjoyable. The landmarks in the sky are the brightest stars and the constellations.

This exercise is designed to familiarize you with several of the major constellations and bright stars that can be seen from mid-northern latitudes on the Earth. Once you know these patterns and can find them in the nighttime sky, you can use them as landmarks or direction finders to locate other stars and constellations and, with practice, you can become an expert at finding your way around the sky.

In history, the constellations were groups of stars, which outlined in crude, formed a familiar object or pattern. In modern astronomy, the constellations are 88 "officially designated" areas of sky, covering the entire celestial sphere with no gaps between them (Section 2-2 of Kaufmann & Freedman, *Universe*, Fifth Ed). However, most people, including most astronomers, think of a constellation more often in terms of the historical star pattern than as a set of invisible boundary lines dividing the sky into sections.

There are other patterns of stars called "asterisms," which are useful to know, even though they are not constellations. Some of these asterisms are part of a larger constellation; for example, the Big Dipper is part of the constellation Ursa Major, or the Big Bear. Other asterisms, such as the Summer Triangle (Figure 2-8, *Universe*), extend over several constellations.

It is important to note that, at any given time of the night, not all of the stars and constellations described in this exercise will be visible. You should be able to find some of them, but others will be below the horizon and therefore hidden from view. To see these constellations, you need to try again at a different time of night (e.g., before dawn) or at a different time of the year. The changing appearance of the night sky is described in Section 2-3 of *Universe*.

Since the purpose of this exercise is for you to become familiar with the nighttime sky, we will first set up the screen to show only stars and constellations, eliminating distractions such as the horizon and daylight, and the Sun, Moon and planets. We will also select a mid-northern latitude on the Earth—New York City, at about 40° north latitude—to allow a clear view of the parts of the sky in which we are most interested, and we will begin by facing north. This flexibility is possible through the magic of a computer!

In this exercise, Part 1 leads you through the steps needed to set up the screen, and Part 2 then guides you to the stars and constellations that can be found by using stars in the Big Dipper as pointers. Some of these alignments are also described in Section 2-3 and Figure 2-6 of *Universe*.

Starting the Program

Start the program by clicking the mouse on the "Starry Night™, Freeman Edition" icon on the "Windows" screen or by running the "Starry Night™ Freeman Edition" program from the location where it was initially installed (e.g., C:\Program Files\Sienna\Starry Night™ Freeman Edition\).

The **initial set-up** of this program is for the "present" time (i.e., Now) at the default location, which is most likely your present location. Each of these parameters can easily be adjusted at any time.

The default function for the mouse is the **Grabber Tool,** whose mouse icon is a hand. The Grabber Tool can be used to change the observer's viewing direction by clicking and holding the left mouse button while the icon is "dragged" anywhere on the screen. To activate the Grabber Tool if it is not already activated, first click Window in the menu bar and look to see if the menu says "Show Floating Palettes" or "Hide Floating Palettes." Click on "Show Floating Palettes" if this is what the menu says, and you should see the icon/location box appear on the screen. You can recognize this box by the four icon buttons in a square at the top of the box. One of these four icons should be the Grabber Tool, which can be activated by clicking on this button. (If the Window menu says "Hide Floating Palettes," then the icon/location box should already be visible on the screen.) Alternatively, you can activate the Grabber Tool by pressing the F4 key.

The Grabber Tool is one of four tools that can be used for various purposes within the program. The other screen tools can be activated by clicking on the appropriate icon in the icon/location box or by using the correct F key. They each have a different mouse icon. The **Object Selection Tool** (click on the arrow icon or press F3) can be used to identify any object on the screen by pointing the arrow icon at the object in question and clicking the left mouse button. The third tool, the **Constellation Selection Tool** (click on the line-shape-and-arrow icon or press F5), can be used to identify the constellation in which the mouse icon has been placed. The fourth tool, the **Magnification Tool** (click on the magnifying glass icon or press F8), can be used to zoom in on a particular region of the sky around the mouse position, the extent of the region being shown by its width in degrees in the box at the bottom of the icon/location box. The magnification can be reversed by clicking the left mouse button while holding down the Control (CTRL) key. The screen icon will change to a minus sign at this time to indicate this "zoom-out" function.

PART 1: Screen Set-up

1. Switch off daylight. You can do this by clicking on Daylight in the Display menu. Or, instead, you can simply press CTRL+D; that is, holding down the CTRL key while you press D.

2. Similarly, switch off the horizon either by clicking on Horizon in the Display menu or by pressing CTRL+H.

3. Switch off the planets by clicking on Planets in the Display menu.

4. Set your location to New York. You can do this by following these steps:
   a. Click on Viewing Location in the Settings menu. A new screen should appear, displaying a map of the world and a list of continents.
   b. Click on the small triangle next to the words "North America" in the list of continents, then click on the triangle next to the words "United States." A list of U.S. cities should appear.
   c. Go to the lower-right-hand corner of the list of cities, and click the down arrow (little black triangle) several times until you can see New York in the list of cities. Click on New York to highlight it.
   d. Click on the Set Location button, which is located to the lower right of the world map.

5. Set the date to 1/01/1999 AD using the following steps:
   a. Go to the Window menu and look to see if Time has a check mark beside it. If it does not, then click Time to activate the date/time box. If there is already a check mark there, then this box should already be activated. You can recognize the date/time box by the button in it, which says "Now." (Alternatively, you can activate the date/time box by pressing CTRL+T.)
   b. There is a row of buttons in the date/time box, one of which has a black square on it. This is the time-stop button. If this button is not already highlighted, then click on it to stop the progress of time.
   c. Today's date is shown in date/time box in the form MONTH/DAY/YEAR AD (e.g., 10/23/1998 AD). Change the date to 1/01/1999 AD. To do this, click on the number corresponding to the MONTH, then type the number 1 (e.g., if the date is 10/23/1998 then click on the 10 and type 1). This will replace the 10 with the number 01. Next, click on the DAY (e.g., 23) and type 1; and finally click on the YEAR (e.g., 1998), and type 1999.

6. Directly below the date in the date/time box, you should see the time in the form HOURS:MINUTES:SECONDS, followed

by AM or PM. Change the time to 12:30:00 PM, just after midday. You can do this the same way that you changed the date in step 5b, above. If the time shown is AM, do not forget to change it to PM (click on AM, then type P).

7. Look near the bottom of the icon/location box and check that the Viewing Elevation is 3 m. If the elevation is any other number of meters, click on the button with the little house on it in this box. The buttons with the rising and falling rockets on them can be used to increase or decrease your elevation; clicking on the house button gets you back to the default value of 3 m.

8. Look again near the bottom of the icon/location box and check that the Field of View is 100°. The plus and minus symbols inside little magnifying glasses can be used to change the field of view, while the third button in the same row resets the field of view to the default value of 100°.

9. Finally, change your viewing location to North. To do this, go to the Window menu and see if Toolbar has a check mark beside it. If it does not, then click on Toolbar. You should now see a row of buttons on the screen, four of which should say N, S, E, and W. Click once on the N button, and you should see the stars move across the screen until you are facing directly north.

   If you are familiar with the Big Dipper, you should now see it near the upper-left-hand corner of the screen.

PART 2: Using the Big Dipper as a Guide to the Sky

1. Find the Big Dipper. This asterism, part of the constellation **Ursa Major**, or the Big Bear, consists of seven moderately bright stars that form the shape of a dipper that you might use to scoop water out of a bucket. It should be on the left-hand or upper-left-hand side of the screen. The bowl of the Big Dipper consists of four stars forming a slightly skewed rectangle, and the handle consists of three stars forming an arc extending toward the left (or upper left) in the orientation on your screen. In the actual sky the Big Dipper can have any orientation, such as upside down or hanging downward from its

handle, depending on the time of night and the time of year. This is because these stars, and indeed all stars, appear to rotate around the North Pole position in our sky.

You can check that you have found the correct pattern of stars by clicking Constellations in the Guides menu to display a set of connecting lines that outline the patterns of stars. You *might* be able to make out the shape of a bear in this, the constellation Ursa Major or the Big Bear! At any rate, you will see this name identifying the group of stars near the upper-left-hand corner of the screen. Click Guides/Constellations a second time to switch the constellations off again; leaving the lines on will create confusion. Since the real sky has no lines in it, it is important to become familiar with finding your way around the sky using only the stars.

2. The Big Dipper is one of the most important constellations to know in the sky because the stars in it can be used as pointers to find many other stars and constellations, particularly the North Pole Star.

   We will need to define two words: "upward" and "downward," relative to the Big Dipper. First tilt your head, or imagine that you have tilted your head, so that the dipper is horizontal with the handle to your left and the bowl opening upward. This defines "upward" relative to the Big Dipper. "Downward" is the opposite direction—the direction in which water would drip if the bowl leaked. For example, if in the real sky the Big Dipper is standing vertically on its handle (bowl above the handle), then upward from the bowl would be toward the left in the sky, and downward from the bowl would be toward the right.

3. Now use the Big Dipper to find the North Star, **Polaris**. To do this, first find the two stars that are on the end of the bowl farthest from the handle. These are often called the "pointer stars." Start from the bottom star (the star at the bottom of the bowl farthest from the handle), and extend an imaginary straight line upward through the top star (at the top of the bowl). If you continue to extend this line upward a total distance approximately equal to 5 times the spacing of these pointer stars, you should find a moderately bright star. This is Polaris, the North Star.

(The North Star is *not* a really bright star—it is actually slightly fainter than the brightest stars in the Big Dipper—but it *is* the brightest star in that little patch of sky.)

If you have found the correct star, then you should see a line of four stars on the screen, curving away from Polaris in an arc toward the handle of the Big Dipper. The first three of these stars are fainter than Polaris and the fourth is about equal in brightness to Polaris. These stars are part of the Little Dipper, which consists of the brightest stars in the constellation Ursa Minor, or the Little Bear.

Click on Guides/Constellations to bring up the constellation lines and names, and check that you have found Ursa Minor correctly; then click the constellation lines off again.

Notice that the two stars in the bowl of the Little Dipper furthest from the North Star (i.e., closest to the Big Dipper's handle) are brighter than all the other stars in the Little Dipper except Polaris. These two stars are often called the "Guardians" because, as the sky turns, they always remain between the Big Dipper and Polaris and "guard" Polaris from the Big Dipper (or from the Great Bear). In a city, with its light pollution, you might find when you look at the real sky that the only stars of the Little Dipper that you can see are Polaris and the two Guardians.

In clicking the constellation lines on and off again, you may find that one of the four stars in the bowl of the Little Dipper is "missing." It is visible in the real sky if the sky is very dark, with no bright moonlight, city lights, or other light pollution. If it is missing from your screen, this is caused by its being fainter than the lowest brightness limit of the stars in the program.

The **brightness**, or "magnitude," of the stars that appear in the sky view in this program can easily be changed to match the expected sky conditions at any site. To do this, simply click on Magnitudes in the Settings menu. A slider control can be adjusted with the mouse to change the star brightness. A particular setting can be tried by using the Apply command. Clicking on OK will then return you to the screen. The stars shown in this program are limited to a sub-set of the stars in the real sky, those that are brighter than a certain limit. Thus, using this "brightness" facility will merely brighten or darken those that are visible but will not bring fainter stars into view.

With the constellation lines turned off, check again that you can find the pointer stars in the Big Dipper and that you can use them to find Polaris. Then check that you can see the Little Dipper, with its two Guardian stars, hanging downward in an arc toward the handle of the Big Dipper.

5. Now, with the constellation lines clicked off, locate the constellation **Cassiopeia** (the Queen of Ethiopia). To do this, start at the handle of the Big Dipper and imagine a straight line drawn from this position through the North Star to approximately an equal distance beyond it. The end of this line should place you in a group of stars shaped like a W. In the orientation on your screen at this time, the right side of the W should be tilted upward. (At other times and in other orientations, it may look more like an M, depending on which side is facing down.)

   Click the constellation lines on to check that you have found Cassiopeia correctly, then click them off again.

   If you look carefully, you may see a faint band of brighter sky on your screen, running downward through Cassiopeia. This band of light depicts the **Milky Way**, made up in the real sky of countless faint stars. Most of these stars are too faint to be seen individually with the unaided eye, but their collective light contributes to a faint glow in the sky.

   With the constellation lines clicked off, check again that you can find Cassiopeia using the handle of the Big Dipper and Polaris.

6. Now find the star **Capella**. To do this, first find the two stars on the top of the bowl of the Big Dipper. One of these two stars marks the point where the handle joins the bowl, and the other is across the top of the bowl from the first star.

   Draw an imaginary straight line through these two stars, starting where the handle joins the bowl, and extend it a long way across the sky (toward the lower right of your screen). The first *really* bright star that you come to is Capella, in the constellation **Auriga**, the Charioteer. (When you look at Capella in the real sky, you may notice that it has a slightly yellowish tint because it is roughly the same temperature as the Sun. However, our Sun is a dwarf star and Capella is a supergiant, so they are very different from each other in size.)

Click the constellation lines on again to check that you have found the constellation Auriga correctly, then click them off again. Capella is the brightest star in Auriga.

You can check that you have located the correct star by using the Object Selection Tool, described above. Click on the arrow icon at the top of the icon/location box, or press the F3 key. The cursor will now be an arrow. Click on the mouse with the arrow adjacent to the chosen star and its name will appear on the screen. To remove this name, click the mouse with the cursor on blank sky. Return to the Grabber Tool by clicking the mouse on the hand icon at the top of the icon/location box, or press the F4 key.

7. To find the next set of stars and constellations, you will have to change your **viewpoint**, or the direction in which you are facing in the sky. To do this, place the cursor hand of the Grabber Tool somewhere on the Big Dipper, then press the left mouse button. KEEP HOLDING THIS BUTTON DOWN and use the mouse to move the hand slowly toward the right across the screen, as explained earlier. The hand drags the sky with it across the screen. Use this action to place the Big Dipper on the right-hand side of the screen, about halfway down from the top of the screen. Make sure that the Big Dipper is close to the right edge of the screen, with all of its stars visible (move the sky back to the left again a little if necessary).

8. Now find the star **Arcturus**. To do this, look at the handle of the Big Dipper. Notice that the handle is curved and forms an arc in the sky. Imagine this arc extended through the sky in the direction away from the bowl (toward the left and somewhat down on your screen). Then you should find that it passes more or less through a bright star near the center of the screen. This star is Arcturus. (A good way to remember how to find this star is, "Follow the arc to Arcturus.")

Use the Object Selection Tool to confirm your identification of Arcturus by clicking on the arrow button in the icon/location box, and then clicking the mouse with the arrow icon placed close to the chosen star. The name of the star will be displayed. Click the mouse again with the icon on blank sky to remove the name, then return to the Grabber Tool.

Arcturus is the brightest star in the constellation **Bootes**, the hunter. (The two o's in Bootes are pronounced separately: "Bo-otes"; to show this, it is often spelled Boötes.) If you look toward the upper right from Arcturus on your screen, you should see a pentagon of five stars with a fainter sixth one just above them. If you ignore the sixth one, then Arcturus plus the other five form a kite-shape or an ice-cream cone in the sky. In the real sky, the whole constellation of Bootes is quite large, and the kite or ice-cream cone forms the most easily seen part of it. The orientation of Bootes in the real sky may be different than on your screen, depending on the time of year and time of night. However, the shape will always be the same, and the brightest star in this constellation can always be found by starting from the Big Dipper and "following the arc to Arcturus."

In the real sky, Arcturus has a yellowish or even slightly orange tint because it is slightly cooler than the Sun.

9. Bootes is hunting the Great Bear—as the sky turns counterclockwise around the North Star, Bootes follows the Great Bear around the sky. If you look below the handle of the Big Dipper on your screen, you should see Bootes' hunting dogs, two relatively faint stars that form the constellation **Canes Venatici** (Latin for "dogs of the hunter"). These two dogs nip at the heels of the Great Bear as it circles the North Star, trying to get away from Bootes.

   Another interesting constellation to find before leaving Bootes is Corona Borealis (the Northern Crown). This is the semicircular arc of stars next to Bootes' shoulder (to the left of the top of the kite or ice-cream cone on your screen).

   Click the constellation lines on to see the full extent of the constellation Bootes and the smaller constellation Canes Venatici, then click them off again.

10. To find the next set of stars, you need to adjust the screen again. The Big Dipper should still be where you set it in step

7, on the right-hand side of the screen about halfway down from the top. The adjustment you need to make is to move the Big Dipper upward, to the upper-right-hand corner of the screen. Do so by following the instructions in step 7. Place the cursor hand on the Big Dipper, hold the left mouse button down, and drag the sky upward. Place the Big Dipper close to the corner, but make sure that all seven stars are still visible.

11. The next star to find is **Spica**. To do so, start with the handle of the Big Dipper and "follow the arc to Arcturus." Then continue this arc to Spica, a relatively bright star. You should find Spica in the bottom left portion of your screen, roughly the same distance past Arcturus that Arcturus is from the handle of the Big Dipper. Spica is considerably hotter than the Sun and has a whitish or even slightly bluish tint when you look at it in the sky.

    Spica is the brightest star in the constellation Virgo, the Virgin. You can check that you have found Virgo correctly by clicking the constellations on and looking for the name "Virgo"; then click them off again. You can also check that you have found the star Spica correctly, using the Object Selection Tool as described above.

12. Now find the star **Regulus**. To do this, go back to the Big Dipper, and use the pointer stars in reverse—instead of pointing upward to find the North Star, point downward from them a distance somewhat greater than the total length of the Big Dipper, or about nine times the spacing of the pointer stars. You should find two moderately bright stars. The one that is farther from the Big Dipper (lower on your screen) and somewhat brighter than the other is the star Regulus, the brightest star in the constellation **Leo**, the Lion. Click the constellations on and off again to check that you have found Leo, and use the Object Location Tool to be certain that you have located Regulus.

    To find the star pattern making up the constellation Leo, first look for a line of stars running from Regulus through the other bright star and forming the shape of a backward question mark. This forms the head and front legs of the lion. Then look for a triangle of stars that forms the hind end and tail of the lion. (Clicking the constellation lines on and off will help you pick out these patterns.) In the sky, the backward question mark and the triangle form the most easily seen parts of the constellation.

Like Spica, Regulus is much hotter than the Sun and has a whitish or even slightly bluish tint when you look at it in the sky.

You have now explored the northern sky, identifying several constellations and bright stars while becoming familiar with the Starry Night™ program and many of its features. If the opportunity arises, you should attempt to explore the real sky and identify these features from your home location. You can use Starry Night™ to show which stars and constellations are visible from your own location at any time by simply placing yourself at your home location and using the Now button in the date/time box, or entering a suitable time such as 10:00 tonight.

If you want to save this configuration for future reference, you can use the "Save as" command in the File menu, giving the configuration a name and saving it in a convenient location. You can then return to it merely by opening this file.

2

Motion of the Sky: Daily Motion Due to Earth's Rotation

The purpose of this exercise is to become familiar with the daily motion of the sky as seen from the Earth, and to see how this motion changes as you move to different latitudes on the Earth. The daily motion of the sky is called diurnal motion, and is discussed in Section 2-3 of Kaufmann and Freedman, *Universe*, Fifth Ed.

The Sun rises in the east, crosses the sky, and sets in the west. However, despite appearances, it is not really the Sun that is moving—the Sun is at rest at the center of the solar system. The apparent daily motion of the Sun is due to our own rotation—the fact that we look out into the Universe in a constantly changing direction as the Earth rotates. Turning toward the Sun causes the Sun to rise above the horizon, and turning away from it again causes it to set below the horizon.

The Sun is just one of a large number of objects in the sky. Other objects visible to the unaided eye are the Moon, five planets (Mercury, Venus, Mars, Jupiter, and Saturn), and many stars. If the daily motion of the sky is due to the Earth's rotation then, as our direction of view swings around the Universe, we should see the entire sky and everything in it constantly rising in the east, passing somewhere overhead, and setting in the west. The Sun should be only one such object.

Also, because the Earth is round, people at different latitudes are standing on ground that is tilted in different directions relative to the universe. Therefore, observers at different latitudes should see the apparent daily motion of the universe from different perspectives.

PART 1: Earth's Rotation and Daily Sky Motion

A. Screen Set-up

1. Restart the program and move to a location at about 45° north latitude. A good example is Seattle, Washington, at about 48° latitude (Settings/Viewing Location/North America/United States/ Seattle/Set Location).

2. Click the time-stop button on the date/time box (if this toolbar is not already on the screen, activate it by clicking Time in the Window menu), then set the date to March 21, 1999 (3/21/1999 AD), and the time to noon (12:00:00 PM). If you do not remember how to perform these steps, refer to Exercise 1, steps 5 and 6.

3. Face south by clicking the S button on the button bar near the top of the screen. If this button bar is not already visible on the screen, activate it by clicking Toolbar in the Window menu. After the screen adjusts direction, you should see a symbol showing south (S) on the horizon near the bottom of your screen.

   You may have expected that the Sun would be directly south at noon, whereas if you chose Seattle for your location the Sun is probably slightly east of south. This small offset is due to the fact that Seattle is not exactly at the center of its time zone (Pacific Time).

4. If the time step in the date/time box does not show 003 minutes, click on the time-step number to highlight it, and type 3. If the time step is not in minutes, click on the little up or down arrows next to the time step to select minutes.

B. Apparent Daily Motion of the Sky Due to the Earth's Rotation

1. First, check the direction of motion of the Sun in the sky by clicking the time-start button, with the right-pointing triangle on it, in the date/time box (the first button to the right of the time-stop button). This will move the time forward in a continuous sequence of three minutes per time step in the present set-up. Compass directions are given along the horizon near the bottom of the screen (if S is centered, then SE and SW should be visible near the left and right sides of the horizon). Motion toward the east would be toward the left on the screen, and motion toward the west would be toward the right. This motion will appear to carry the Sun across the sky in an arc, reaching its highest point around midday.

   After noting the direction of motion of the Sun in the sky, click the time-stop button and reset the time to noon (12:00:00 PM).

2. On your screen, the blue sky, which is produced by scattering of sunlight by the Earth's atmosphere, hides almost everything but the Sun. The Moon is bright enough to be seen in the daytime, and sometimes Venus is visible if you know exactly where to look. All other astronomical objects are "lost" in the brightness of the blue sky. In reality, however, all astronomical objects that are above the horizon are actually there in the daytime. Through the magic of computers, you can make all of the other objects visible with the click of a switch! Switch off daylight by clicking on Daylight in the Display menu. This shows the bright Sun, but eliminates the blue sky so you can see everything else. (This is the same view that you would have if the Earth had no atmosphere—the Sun in a dark sky with all of the other objects shining as if it were nighttime. Astronauts get an equivalent view to this from orbit or from the surface of the Moon.)

   Now click the time-start button again and observe what happens. If the motion of the Sun across the sky were due to the Sun actually moving, then the rest of the sky would remain stationary as the Sun moved past it. If the motion of the Sun were an illusion due entirely to our own motion, then our motion would affect all objects equally and the entire sky would appear to move. As you see, the Sun and all other objects follow arcs across the sky from east to west together.

When you have tried this step, click the time-stop button, reset the time to noon (12:00:00 PM), and click daylight back on.

3. Now try estimating the speed of motion of the Sun in the sky. Fold a sheet of paper into a strip so you can hold it close to the screen and still see the time-step buttons in the date/time box. (Do this very carefully to prevent any damage to the screen. If you are using a laptop, you may wish to skip this step because of the softer screen or find a safer way to measure distances on the screen.) Make a mark on the paper, and place the paper horizontally across the screen with the mark next to the Sun. Now, holding the paper so it does not move, click the time-start button and let the Sun move until the time shows 1:00:00 PM. Stop the motion at 1:00:00 PM exactly. (You can back the motion up or move it ahead one step at a time by using the single-step buttons—the outside two buttons in the date/time box—if you need to.)

   Now make a second mark on the paper at the new location of the Sun. Remove the paper from the screen and measure the distance between the marks.

   Next, measure the distance between S and SW along the horizon on the screen. This distance corresponds to a 45° angle (from S to W would be 90°, and S to SW is half this).

   Finally, calculate the angle that the Sun moved in one hour by dividing the distance between the marks on the paper by the distance between S and SW, and multiplying the result by 45°. (For example, if the distance between the marks were 1/5 of the distance between S and SW, then the Sun would have moved $1/5 \times 45° = 9°$ in one hour. Your answer should be larger than this.)

   Now, the Earth rotates through a full circle in 24 hours, and there are 360° in a circle. Therefore, through how many degrees does the Earth rotate in one hour? The answer you get should equal your measurement of the motion of the Sun in one hour, in degrees. In fact the two answers will probably differ by a small amount because of uncertainties in measurement and small but unavoidable distortions arising from mapping a spherical sky onto a flat screen. But the two answers should agree reasonably closely.

PART 2: Eastern Rising of Objects in the Sky

In this part of the exercise, you can see how the Sun and other objects behave as they rise above the eastern horizon.

A. Screen Set-up

Check that the sky motion is stopped. If the sky is moving, click the time-stop button in the date/time box. Then set the time to 6:00:00 AM, and click the E button in the button bar near the top of the screen, to face east. (If this button bar is not already visible on the screen, activate it by clicking Toolbar in the Window menu).

It is also important for this step to have a straight horizon. If the horizon is curved up or down, then place the cursor hand on the horizon, hold down the left mouse button, and drag the horizon up or down until it is straight.

B. Measuring the Angle that the Sun's Track Makes to the Horizon at Sunrise

Carefully note the point on the horizon where the Sun rises. If necessary, use the single-step buttons in the date/time box to move time forward or backward one step at a time, to place the Sun exactly on the horizon.

Click the time-start button and let the Sun rise until it reaches the edge of the screen. At that point, click the time-stop button. If necessary, use the single-step buttons to bring the Sun back onto the screen or advance it until it reaches the edge of the screen. Use a piece of paper laid on the screen to mark the angle that the Sun track makes with the horizon. Alternatively, if you have a protractor, then you can measure this angle in degrees. Work gently and be careful not to damage the screen when doing these measurements. (Again, if you are using a laptop, be especially careful or make hand-drawn diagrams without measuring directly on the screen.) The angle should be around 40° to 50° if the latitude of your location is between 40° and 50°.

Write down the latitude of your location, and the angle of rising of the Sun. The latitude and longitude are printed in the icon/location box; if this box is not visible on the screen, you can activate it by clicking Show Floating Palettes in the Window menu.

Set the clock back to 6:00:00 AM, set your location to Honolulu (Settings/Viewing Location/North America/United States/Honolulu/Set Location), and follow a similar procedure to measure the new angle of rising of the Sun, this time allowing the Sun to reach the upper edge of the screen. Write down the latitude and the rising angle for Honolulu. Repeat this step for the following viewing locations, resetting the clock each time and writing down the latitude and the rising angle for each location:

Quito, Ecuador	(South America/Others/Quito, Ecuador)
Dunedin, New Zealand	(Australasia/New Zealand/Dunedin)
Murmansk, Russia	(Europe/Russia/Murmansk)
The North Pole	(click on the latitude number in the icon/location box, and type 90)

Questions

1. At what angle does the Sun rise for someone at 0° latitude?

2. At what angle does the Sun rise for someone at 90° latitude? (If the Sun just skims along the horizon, then the rising angle is 0°.)

3. What happens to the rising angle of the Sun as you move from the equator (0°) to the North Pole (90°).

4. Based on your answers to the questions above, which one of the following statements do you think is correct?
   a. The rising angle of the Sun is equal to your latitude.
   b. The rising angle of the Sun is equal to 90° minus your latitude.
   c. The rising angle of the Sun does not depend on your latitude.

5. What is different about the sunrise in Dunedin, New Zealand compared with that at Seattle, Washington?

6. Suppose that you are in Dunedin, New Zealand, the time is noon, and you are facing the point on the horizon where the Sun rose. Based on how the Sun rises in Dunedin, in which direction (left or right) would you need to turn to face the Sun at noon in Dunedin? How does this compare to Seattle?

7. Based on how the Sun rises in Dunedin, in which compass direction (north or south) would you need to face to see the Sun at noon? How does this compare to Seattle?

8. Based on how the Sun rises in Dunedin, in which direction would the Sun be moving at noon (from left to right or from right to left)? If you wish, you can check this directly by setting the time to 12:00:00 PM and your location to Dunedin, and clicking the time-start button. How does this compare to Seattle?

9. In which compass direction would the Sun be moving at noon in Dunedin (from east to west or from west to east)? Check the compass directions near the bottom of the screen to find if you are right. How does this compare to Seattle, Washington?

C. Rising of Other Objects

Check that the sky motion has been stopped, and if not then click the time-stop button. Set your location back to Seattle, set the time to 1:00:00 AM, set the compass direction to east (click E on the button bar near the top of the screen; if necessary, first click Window/Toolbar to activate this button bar) and adjust the horizon to be horizontal again. Then click the time-start button. You should see stars continuously rising above the eastern horizon at the same angle from which the Sun rose. Throughout the night, new stars and constellations are always rising in the east, and the old ones are setting in the west and disappearing from view.

Now take a moment to see what happens as objects set. Click the time-stop button, then click W (for west) at the top of the screen, and set the time to 4:00:00 PM. If the sky is not blue, then switch on daylight by clicking Daylight in the Display menu.

Click the time-start button, and watch what happens until about 9 PM. You should see the Sun approach the horizon and set at the same angle from which it rose in the east. After the sky gets dark you should see all of the stars setting at the same angle as the Sun set. This is a further indication that the motion of the sky is an illusion caused by our own rotation. As an added bonus, Venus and the Moon are seen to set just after the Sun on this date and from this location.

D. The Effect of Latitude on the Appearance of Constellations: Southward View

Latitude on the Earth not only affects the angle from which we view the Sun but it also determines whether (and how) we see all of the other objects in the sky. In this step you can see how our latitude affects the visibility of the constellations that we see when we look southward (from the northern hemisphere). Later in this exercise, you can see how their visibility changes when we look northward from different locations in the northern hemisphere.

Check that the sky motion is stopped, and if not then click the time-stop button. Set your location to Kingston, Jamaica (click Settings/Viewing Location/Central America/Kingston, Jamaica/Set Location), and click the S button to face south. Then set the date to June 10, 1999 (06/10/1999 AD) and the time to midnight (12:00:00 AM).

Find the constellation Scorpius. It should be in the upper middle part of the screen, and it is shaped like a fishhook. If you do not see it, or wish to check your identification, switch on the constellation lines and names by clicking Constellations in the Guides menu, look for Scorpius, then click Guides/Constellations again to switch the constellations off again.

Notice carefully how far Scorpius is above the southern horizon. Also notice that you are seeing the constellation due south, and therefore you are seeing it at the highest position it reaches in the sky for that location.

Now set your location to each of the cities listed on the following page. In each case notice where Scorpius is located relative to the horizon. For Calgary and Fairbanks, you will have to reset the time to 12:00:00 AM, since they are in different time zones than Kingston or New York. You may also have to check the date. When you are viewing Scorpius from Calgary, note carefully where Scorpius is relative to Libra, so you may find Scorpius again when you move to Fairbanks. At any time click the constellations on and off again if it helps in finding Scorpius.

New York	(North America/United States/New York)
Calgary, Alberta	(North America/Canada/Calgary)
Fairbanks, Alaska	(North America/United States/Fairbanks, AK)

(*Note:* You may have to turn off the daylight to see the stars in Fairbanks, since the sky never gets completely dark in the summertime at this high latitude. To do this, click Daylight in the Display menu.

Questions

1. For which of these cities is the pattern of stars in Scorpius completely above the horizon?

2. At about what latitude does the tail of Scorpius begin to disappear from view (i.e., never rises above the horizon)?

3. Above about what latitude is *no* part of Scorpius ever visible?

4. Suppose that a scientist in Fairbanks has applied for a research grant to study a globular cluster of stars near the middle of Scorpius, using an observatory just outside Fairbanks. If you were on the granting agency, what would be your response to this proposal? Would your response change if the research proposal included a request for funds to travel to an observatory in Texas?

E. The Effect of Latitude on the Appearance of Constellations: Northward View

In this part of the exercise, you can see how the visibility of constellations in the sky changes when we look northward from different locations in the **northern** hemisphere.

Check that the sky motion is stopped; if not, click the time-stop button. Set your location to Minneapolis, Minnesota (click Settings/Viewing

Location/North America/United States/Minneapolis/Set Location), and click the N button to face north. Set the date to October 1, 1999 (10/01/1999 AD) and the time to 7 PM (7:00:00 PM). Switch off daylight by clicking the button with the small sun on it.

Find the Big Dipper in the left part of the screen, and use it to find the North Star, Polaris.

View Northward from Minneapolis

Write down the latitude of Minneapolis, and make a note of where the North Star is, compared to the top and bottom of the screen (e.g., close to the top, about half way down, close to the bottom, etc.).

Click the time-start button in the date/time box, and watch what happens in the sky. Stars on the left, including the Big Dipper, are moving downward, whereas stars on the right, including Cassiopeia, are moving upward. Is there one star that remains at rest while all other stars move around it in circles? Which star is this?

As the sky rotates around the pole, watch what happens to the Big Dipper, especially between about 10 PM and 1 AM. Notice that, on the lower-left-hand side of the screen, stars are setting below the horizon, and on the lower-right-hand side of the screen stars are rising above the horizon; but, as seen from Minneapolis, the Big Dipper never sets. It approaches the northern horizon, but passes above the horizon without setting, and then gets higher in the sky again. Stars or constellations that move in circles around the Pole without ever setting are called circumpolar.

View Northward from Houston

Stop the sky motion by clicking the time-stop button, set your location to Houston, TX (click Settings/Viewing Location/North America/United States/Houston/Set Location), and then set the time to 7:00:00 PM on October 1, 1999 (10/01/1999 AD).

Write down the latitude of Houston, and make a note of where the North Star is, compared to the top and bottom of the screen (e.g., close to the top, about half way down, close to the bottom, etc.).

Click the time-start button and watch the motion of the sky. Notice that all stars of the Big Dipper, except the one that is closest to the North Star, set below the horizon.

View Northward from Port of Spain, Trinidad

Click time-stop, set your location to Port of Spain, Trinidad (click Settings/Viewing Location/Central America/Port of Spain/Trinidad/Set Location), and then set the time to 4:00:00 PM on 10/01/1999 AD.

Write down the latitude of Port of Spain, and make a note of where the North Star is, compared to the top and bottom of the screen (e.g., close to the top, about half way down, close to the bottom, etc.).

Click the time-start button and watch the motion of the sky. What happens to the Big Dipper?

View Northward from Fairbanks, Alaska

Finally, click time-stop, set your location to Fairbanks, Alaska (click Settings/Viewing Location/North America/United States/Fairbanks, AK/Set Location), and set the time to 8:00:00 PM on 10/01/1999 AD.

Write down the latitude of Fairbanks, and make a note of where the North Star is, compared to the top and bottom of the screen (e.g., close to the top, about half way down, close to the bottom, etc.). If you cannot see the North Star, use the Big Dipper to estimate its location.

Click the time-start button and watch the motion of the sky. How does the number of circumpolar stars for Fairbanks compare to the number of circumpolar stars for Port of Spain?

Questions

1. Based on your notes about latitude and the location of the North Star above the horizon, which one of the following statements do you think is correct?
   a. The angle of the North Star above the horizon equals your latitude.
   b. The angle of the North Star above the horizon equals 90° minus your latitude.
   c. The angle of the North Star above the horizon does not depend on your latitude.

2. Where would you expect to see the North Star if you were standing at the North Pole?

3. Where would you expect to see the North Star if you were standing on the equator? (You might like to check this one by setting your location to Quito, Ecuador, in South America, and the time equal to 3:00:00 PM on 10/01/1999. Use the Big Dipper to find the North Star.)

4. If you were south of the equator, say in Australia, would you expect to see circumpolar constellations anywhere in the sky? If so, in what part of the sky would they be?

Answers

Measuring the Angle That the Sun's Track Makes to the Horizon at Sunrise

1. 90° from the horizon (i.e., the stars rise straight up from the horizon).

2. 0°.

3. The angle gets smaller (the stars rise at an angle closer to the horizon).

4. B. The rising angle of the Sun is equal to 90° minus your latitude.

5. At both locations the Sun rises in the East, but in Dunedin, New Zealand, the Sun's path angles toward the North after sunrise, whereas in Seattle, Washington, the path angles toward the south.

6. You would have to turn toward the left in Dunedin, but toward the right in Seattle.

7. You would need to face North in Dunedin, south in Seattle.

8. From right to left in Dunedin, from left to right in Seattle.

9. From east to west in both Dunedin and Seattle.

The Effect of Latitude on the Appearance of Constellations: Southward View

1. Kingston, Jamaica, and New York.

2. Between latitudes 41° (New York) and 51° (Calgary); so perhaps about 45°. (You can test this by clicking on the latitude number in the icon/location box with the date and time set as stated for part D, then typing a series of latitudes from 41° to 51°, one after the other. You will see Scorpius approaching and then sinking below the horizon as you increase the latitude of your viewing location.)

3. Probably about 70°, since just a few stars of Scorpius are visible above the horizon in Fairbanks, at latitude 65°. (You can test this the same way as given in the answer to question 2, above.)

4. Deny the grant; a globular cluster in the middle of Scorpius would never be visible from the observatory because it would never rise above the horizon. Yes, approve the grant, because Scorpius is easily visible from Texas.

The Effect of Latitude on the Appearance of Constellations: Northward View

1. A. The angle of the North Star above the horizon equals your latitude.

2. Directly overhead (at the zenith).

3. Exactly on the horizon (actually it would appear to be slightly above the horizon because of atmospheric refraction).

4. Yes, in the southern part of the sky, around the South Celestial Pole.

The Seasons

3

In this exercise you can learn how the times and locations of sunrise and sunset change with the seasons, and how the seasonal changes vary with latitude on the Earth. The reason why the Earth has seasons is discussed in Section 2-5 of Kaufmann and Freedman, *Universe*, Fifth Ed.

PART 1: Changes in the Position and Time of Sunrise Through One Year

A. Screen Set-up

1. Start by setting your location to New York City.

2. Click the time-stop button on the date/time box (if this toolbar is not already on the screen, activate it by clicking Time in the Window menu).

3. Face east by clicking E on the button bar near the top of the screen. If this button bar is not already visible on the screen, activate it by clicking Toolbar in the Window menu. After the screen adjusts direction, you should see a symbol showing east (E) on the horizon near the bottom of your screen.

   If the horizon is curved, then straighten it by placing the cursor hand on the screen, pressing the left mouse button, and dragging the screen straight upward or downward until the horizon is straight.

4. Notice that there is a little sun symbol in the date/time box to the left of the digital time readout. This sun symbol is the standard-time/daylight-time button and denotes daylight time if the sun is yellow or standard time if the sun is not yellow.

   The time should be set to standard time (sun symbol not colored yellow). If the sun symbol is yellow then click on the symbol once to set the time to standard time. (Alternatively, click on the sun symbol a few times and choose the setting corresponding to the earlier time in the readout.) We will maintain standard time all year so that the various trends show up more clearly (e.g., sunrise getting earlier as the months go by).

5. Set the date to March 21, 1999 (3/21/1999 AD) and the time to 6:00:00 AM.

6. If the time step in the date/time box does not show 003 minutes, click on the time-step number to highlight it, and type 3. If the time-step units are not minutes, click on the time-step units to highlight them, and click on the little up or down arrows to select minutes.

7. Use the single-step buttons in the date/time box (the outermost buttons in this box, marked by arrows with a short, vertical line through them) to place the middle of the Sun exactly on the horizon. This step should be done carefully, since the Sun is depicted on the screen as a small yellow disk that is somewhat difficult to see, particularly if a tree is in the way. If you prefer, you can remove the trees from the screen using the following steps: click Options in the Settings menu, click the down arrow to see a list of options, click Horizon, click off the Scenery, click OK to return to the screen.

   The location of the Sun on the horizon should be at or very close to the east point on the horizon, denoted by E. If this is not the case, check that the date is March 21.

B. Procedure

1. Place a strip of paper about 3 cm in height across the horizon on your screen and attach it with tape to the edge of the monitor (*do not* put tape on the screen itself). Mark the positions

SE, E, and NE on the paper. Be careful not to damage your screen as you do this. On this strip mark where the Sun rises on March 21, and write the month (March) and the time of sunrise above or below the mark.

2. Change the month to April while keeping the day and year fixed (i.e., 4/21/1999). Notice that at 6 am on April 21 the Sun has already risen and is up in the sky; therefore, run time backward for a while to place the center of the Sun exactly on the horizon again.

   On your paper strip along the horizon, mark the location of sunrise on April 21, and write "April" and the new time of sunrise above or below the mark.

3. Keeping the date set to the 21st of the month, repeat the above steps for May, June, July, August, September, October, November, December, January, and February, using a new line to mark the positions, dates, and sunrise times for the return portion of the path. You can change the year to 2000 in January, or leave it at 1999.

4. To represent the movement of the sunrise position graphically, use your measurements to plot the sunrise position as a function of time. A simple way to do this is to slide your strip of paper downward over a sheet of graph paper, plotting the sunrise positions horizontally for each month in consecutive order (i.e., March, April, May, . . .), placing the months at equal vertical intervals on your graph paper. Be careful to keep one edge of your strip aligned with a vertical reference line on the graph paper.

5. Plot a second graph of the sunrise time against time of the year. Choose the scale for sunrise time carefully to include 60 intervals per hour, for example, select 1 cm = 10 minutes of time.

Questions

1. What is the shape of the graph showing sunrise position as a function of time of the year? (E.g., linear change with time, sine wave change with time, . . .)

2. For which months of the year does the Sun rise exactly (or almost exactly) due east?

3. For which months does it rise to the north of east? During these months, does it rise earlier or later than it does on March 21?

4. For which points does it rise to the south of east? During these months, does it rise earlier or later than it does on March 21?

5. In which month does the Sun rise furthest north? In which month does it rise furthest south? Which of these months corresponds to the latest sunrise? Earliest sunrise?

6. During which months is the Sun moving northward (sunrise further north than it was the previous month)? During which months is it moving southward? What happens to the time of sunrise as the Sun moves northward? Southward?

7. Use your graph to estimate when the sunrise position is changing most rapidly.

8. Are the answers to the above questions exactly as you would have expected? If not, describe what was new or surprising for you.

9. You can now use your graph to tell where on the New York horizon the Sun will rise on any day of the year, and estimate the sunrise time. Remember that the graph is plotted for the 21st of each month. Estimate the time of sunrise for November 5, which is 1/2 month after October 21.

PART 2: The Length of the Day (Sunrise to Sunset)

A. Screen Set-up

Reset the screen as follows:

viewing direction, west (W);
horizon straight (drag the screen upward or downward if necessary);
date, March 21, 1999;
time, 5:45:00 PM;
time step, 3 minutes.

B. Procedure

1. Run time forward a step at a time to place the setting Sun exactly on the horizon. Write down the date and time of sunset.

2. Repeat for June 21, 1999, September 1, 1999, and December 1, 1999. Then use the times of sunrise and sunset on these dates to calculate the length of the day in New York City for all four dates.

Questions

1. What is the length of the day in New York City on March 21 (near the start of Spring) and September 21 (near the start of Fall)?

2. What is the length of the day in New York City on June 21 (near the start of Summer)? What is the length of the night on June 21?

3. Repeat the previous question for December 21 (near the start of Winter). What approximate relationship do you notice between the lengths of day and night for June 21 and December 21? (E.g., How does the length of the day on June 21 compare to the length of the night on December 21?)

4. What happens to the length of the day as the rising or setting point of the Sun moves northward along the horizon? Southward?

5. We have kept the time set to standard time through this exercise, but in fact most areas use daylight savings time (or daylight time) between April and September. If daylight time were in effect, would it have any effect on the length of the day or night? (Remember that we set our clocks one hour ahead when daylight time begins, so *both* sunrise and sunset take place one hour later than they would if we had left the clocks at standard time.)

PART 3: Lattitude Dependence of Sunrise Time and Location

A. Screen Set-up

1. For this part of the exercise, you need to be able to see the Sun even when it is below the horizon. To do this, click Options in the Settings menu, then click Horizon type, See through, OK. (The magic of computers again!)

2. Set your location to Quito, Ecuador, almost exactly on the equator (Settings/Viewing Location/South America/Others/Quito, Ecuador/Set Location).

3. Set the date to December 21, 1999 and the time to 6:12:00 AM.

4. Click E to face east (you will see that we have chosen sunrise for this location and date), and drag the screen upward to straighten the horizon.

5. Click Equatorial Grid in the Guides menu (or press CTRL+2) to switch on the equatorial coordinate system.

The lines that rise vertically from the horizon are **declination** lines (celestial latitude) and the lines that run from left to right are **right ascension** lines (celestial longitude). (The declination lines curve outward at the top of the screen because of distortions introduced by mapping a spherical sky onto a flat screen. In the real sky, the declination lines are parallel circles that decrease in size toward the north and south poles, just as the lines of latitude on a globe of the Earth are parallel circles of decreasing size.)

The declination line running straight up from E on the horizon is the **celestial equator** (0° declination), and any two adjacent declination lines are 10° apart. Thus the first line to the left of the celestial equator is 10° N declination, the next one to the left is 20° N declination, and so on.

In an equivalent way, the lines of right ascension (RA) are 1 hour apart, each 1-hour interval in RA being the angle through which the Earth rotates in 1 hour or, from an Earth-bound viewer, the angle through which a star moves in our sky in 1 hour.

B. Procedure

When we move northward or southward, we are moving over the curved surface of the Earth. This causes our horizon to tilt relative to the fixed sky. However, as we tilt, our "straight up" direction tilts along with us. Consequently our horizon always appears horizontal as seen by us, and we see the sky (apparently) tilting relative to our horizon. This tilt of the sky relative to our horizon causes the location and time of sunrise to change as we change our latitude on the Earth. This part of the exercise allows you to see how this happens.

Note that there is a pair of right ascension and declination lines which cross almost exactly at the E on the horizon. In the following steps, this crossing point will remain almost exactly fixed as long as the time remains fixed, even though the sky rotates as you change your latitude. This gives you a fixed reference point.

1. If you count the declination lines between the Sun and the celestial equator (marked by E on the horizon), then you can see that the declination of the Sun on December 21 is 23 1/2° (2 1/3 declination lines) south of the celestial equator.

   Notice that, because the declination lines are perpendicular to the horizon, the declination lines also mark the horizon off in 10° intervals. Therefore, for a person in Quito, or anywhere else on the Earth's equator, the Sun rises at a point 23 1/2° south of the East point on the horizon on December 21.

2. Note carefully where the Sun is located on the horizon at sunrise for Quito (e.g., is it located behind a particular tree, just to the right of a particular tree, etc.?). Drawing a quick diagram or using a strip of paper again might help you remember this location.

3. Now we are going to move northward on the Earth's surface from the equator, keeping our longitude constant. Look in the icon/location box (if this box is not visible on the screen you can activate it by clicking Show Floating Palettes in the Window menu), and find where it states that your current latitude is 0° (for Quito, Ecuador, this is printed as 0 S, because Quito is a fraction of a degree south of the equator). Click once on the 0 to highlight it.

Now type 5 to set your latitude to 5° N. Notice that the declination lines now tilt downward on the right, placing the Sun below the horizon. Thus at the same instant of time when the Sun is rising for someone on the equator, it is still below the horizon for someone at 5° N latitude. Consequently, in December the Sun will rise later for a person at 5° N latitude than for a person (at the same longitude) at the equator.

4. Now type 10 to change your latitude to 10° N. (If you have not touched the mouse button then the latitude number should still be highlighted; if it is not, then click on the latitude number to highlight it again before typing 10.) Watch what happens to the tilt of the declination lines and to the distance of the Sun below the horizon.

   Repeat this step for 15, 20, 30, 40, and 50 degrees north latitude, each time watching what happens to the tilt of the declination lines and to the distance of the Sun below the horizon.

   Notice that, as you increase your latitude, the Sun moves downward in an arc of a circle. This is because the Sun is a constant 23 1/2° from the fixed reference point (the crossing point at the E, described above). However, the declination lines are straight; therefore, at latitude 50° N, the declination lines near the Sun (20° S and 30° S declination) intersect the horizon considerably to the right (or south around the horizon) from the point where the Sun was on the horizon at Quito.

5. Now click the time-start button, let the Sun approach the horizon, and click the time-stop button when the Sun is exactly on the horizon (just rising). Use the single-step buttons to adjust the time, if needed. Notice that the path of the Sun on the screen is parallel to the declination lines, and therefore the Sun rises considerably to the right (south) of where it rose in Quito.

   Thus you can see that in winter in the northern hemisphere observers further north see the Sun rise later and at a point further south around the horizon. You can also see that this is a geometric effect caused by the increasing tilt of the celestial sphere relative to our horizon (or really the increasing tilt of our horizon relative to the celestial sphere!) as we move further north.

6. This northern movement across the Earth can be taken to extremes! Try moving to 67°N latitude. You need to move the sky so that you are viewing to the South. Let time run until the Sun reaches the South. You will see that the Sun barely reaches the Horizon on this date at this latitude, which is just above the Arctic Circle. North of this latitude is the arctic, where the Sun stays below the horizon for at least 1 day per year. The further north you go, the longer the period of 24-hour darkness.

7. The extreme position is, of course, the **North Pole of Earth**, at 90°N latitude. Move to this latitude and adjust the date to March 21, 1999 (3/21/1999 AD). You will see that the Sun is just above the horizon. If you run time forward, you will find that the Sun remains just above the horizon and tracks around it as time progresses. To experience what this must be like in real life, try magnifying the view of the Sun by clicking on the Magnification Tool, in the icon/location box (or press F8), until the field of view is 14°, say. The Sun just sails majestically along the horizon! Of course, there would be no trees at the North Pole, just icefloes!

   In the winter at the North Pole, between about September 21 and March 21, the Sun will always be below the horizon.

Answers

Position and Time of Sunrise

1. Sine wave.

2. March and September.

3. April through to August. Earlier.

4. October through to February. Later.

5. June. December. December. June.

6. January to May. July to November. Sunrise gets earlier. Sunrise gets later.

7. March and September.

8. About 6:35 AM.

Length of the Day

1. About 12 hours.

2. Length of day, 15 hours; length of night, 9 hours.

3. Length of day, 9 hours; length of night, 15 hours. The length of the day on June 21 is almost exactly equal to the length of the night on December 21 and the length of the night on June 21 is almost exactly equal to the length of the day on December 21.

4. Rising point moving northward: days get longer; rising point moving southward: days get shorter.

5. The length of the day is not affected by daylight savings time; since both sunrise and sunset take place one hour later than they would if we had left the clocks at standard time, the time interval between them has not changed.

The Moon's Motion and Phases

The Moon orbits the Earth in a period with respect to the stars (its **sidereal period**) of 27.3 days. It is illuminated by the Sun and, when viewed from the Earth at different positions in this orbit, shows the various **phases**. The cycle of lunar phases is described in Section 3-1 of Kaufmann and Freedman, *Universe*, Fifth Ed.

In this monthly cycle, the Moon appears first as a very thin crescent in the western sky, quite close to the Sun. It then proceeds to show more of its illuminated side to Earth as it moves farther from the Sun in the sky, passing in this **waxing**, or growing, half of the cycle through **crescent, first quarter**, and **gibbous** phases before reaching **full moon**. At this point, the Moon is opposite the Sun in our sky. Its distance from the Sun, measured across the sky, then begins to decrease, and it enters the **waning**, or diminishing, half of the cycle, proceeding through **gibbous**, **third quarter**, and **crescent** phases before reaching new moon again.

Because of the orbital motion of the Earth–Moon system around the Sun, the period between successive full moons (its **synodic period**) is longer than the Moon's actual orbital period by more than 2 days, a full lunar phase cycle taking 29.5 days. This effect is illustrated in Figure 3-5 of *Universe*.

In this exercise, we will demonstrate the motion of the Moon against the background stars and estimate its angular speed. We will also show the phases of the Moon and relate them to the position of the Moon with respect to the Sun in the sky.

PART 1: Moon's Motion Across the Sky and its Angular Speed

A. Screen Set-up

1. After starting the program, check the icon/location box to see that the field of view is 100° (if this box is not visible on the screen, you can activate it by clicking Show Floating Palettes in the Window menu). If the field of view is not 100°, then click on the house button to reset the field of view to 100°.

2. Set the location to Chicago (Settings/Viewing Location/North America/United States/Chicago, IL/Set Location).

3. If your viewing direction is not already toward the south (S), then click the S button on the button bar near the top of the screen. If this button bar is not visible, activate it by clicking Toolbar in the Window menu. After the screen adjusts direction, you should see the symbol S on the horizon near the bottom of your screen.

4. Stop the time by clicking the time-stop button in the date/time box (if this box is not visible on the screen, activate it by clicking Time in the Window menu); then
   a. set the date to July 27, 1999 (7/27/1999 AD);
   b. set the time to 11:45:00 PM; at this time, the Moon is close to its full phase;
   c. set the time interval in the date/time box to 1 sidereal day.

   Using a time interval of one sidereal day means that the background sky remains stationary as time is changed by single time steps.

5. Check that the Daylight Saving Time option is turned off. To do this, check the small sun icon to the left of the time in the date/time box. The sun icon is yellow with 8 rays if DST is activated, but background color with 4 rays if DST is not activated. If the icon is yellow, then click on it to deactivate DST.

6. The program can show the Moon as a small dot or as an enlarged image. For the initial run of this exercise, ensure that the enlarged image is activated. (Click Settings, then Preferences; if there is no check mark beside Enlarge Moon Size, then click on Enlarge Moon Size.)

B. Procedure

1. Because of the Earth's rotation, the whole sky, including the Moon, appears to move westward as time progresses. However, the Moon also moves independently because of its orbital motion around the Earth. We can see this motion of the Moon against the background stars if we step time in 1 sidereal day intervals, because then the background stars return to the same position night-by-night, but Moon will move past these stars from one night to the next.

   Demonstrate this motion by advancing time by several sidereal day steps. Note the direction of the Moon's motion. (It looks as though the stars do not move, but in fact the computer advances them through exactly one complete "rotation" around the Earth each time you click the button.)

2. To measure the Moon's angular speed against the background stars, we can change the date by sidereal-day steps and estimate how many days it takes the Moon to move across the full 100° width of the screen. To do this, step time backward in sidereal day steps until the date is July 24, 1999 (7/24/1999), at which point the time will be 12:00:44 AM. The Moon should be near the right-hand edge of the screen. Use the Grabber Tool to slide the sky over toward the right (to the West) until the Moon lies right at the edge of the screen.

   Advance the time in 1 sidereal day steps and count how many days it takes the Moon to reach the left edge of the screen, estimating the final fraction of a day that would be needed in order to just reach the edge. After each click of the button, the sky has "rotated" once around the Earth (or actually the Earth rotates once) until the stars are in exactly the same position as they were before the click. For this reason, the stars do not seem to move. However, the Moon moves in its orbit during that time, and this motion of the Moon in one sidereal day is visible against the background stars. Estimate the speed of the Moon by dividing 100° by the number of days it took to move across the 100° screen.

3. It is of interest to track the Moon s motion with respect to the **ecliptic**, which is the path of the Sun against the background stars. Switch on the ecliptic (click Ecliptic in the Guides menu, or press CTRL+3), and step time backward so that the Moon retraces its path back to the right edge of the screen. The Moon's orbit is inclined at about 5° to the ecliptic plane, hence the Moon is seen to follow a path across the sky that is close to the ecliptic, but does not exactly follow it. The Moon's path crosses the ecliptic plane at two points, known as the **nodes** of its orbit, one of which should be visible in the part of the path you have watched on the screen. To display the Moon's orbit, click Planets in the Window menu to display the planet box, and look to see if there is a list of planets; if not, click on the triangle next to the Sun to display the planet names. Then look to see if the Moon is listed underneath the Earth; if not, click on the triangle to the left of the Earth to display the Moon. Finally, click on the "Orbit Column" next to the Moon. Clicking here again will remove the Moon's orbit from the screen. Clicking on the X at the top of the box will remove the box from the screen. The orbit of the Moon is described in Section 3-2 and Figure 3-6 of *Universe.*

Questions

1. In which direction does the Moon move night-by-night against the background stars?
2. At what speed does it move across the sky?
3. What path does the Moon take (curved, straight, along the ecliptic . . .)?

PART 2. The Phases of the Moon

To demonstrate the phases of the Moon, we must observe the sky at different times of the night on different days. For example, the best time to observe the waxing phases is just after sunset, when the Moon appears in the western sky. The phases around full moon are best observed near midnight, while the waning phases can be most effectively seen in the dawn sky.

A. Waxing Phases

Screen Set-up

1. Check that the location is Chicago, IL, USA.
2. Set view direction to W.
3. Set the date to July 13, 1999 (7/13/1999 AD).
4. Check that Daylight Saving Time is OFF.
5. Set the time to 7:30:00 PM, just after sunset on this day.
6. Check that Daylight is ON. If it is not, click on Daylight in the Display menu.
7. Set the time step to 1 solar day (the unit should be days, not sidereal days, in time-step box).
8. Check that "Enlarge Moon Image" is still activated. This gives a false view of the size of the Moon but is illustrative of the movement and phase of the Moon during this part of the exercise.

Procedure

If your set-up is correct, you will see the Sun just below the horizon and the Moon appearing as a faint disk just to the east of it, in the twilight. You can see that the unilluminated side of the Moon is visible at this time, which is why the Moon looks like a faint disk and not just as a thin crescent. This is a result of earthshine, the effect of sunlight reflected from the surface of the Earth illuminating the Moon. This was referred to in earlier times as "the old moon in the new moon's arms."

Advance the time by 1-day steps, and watch the motion of the Moon away from the Sun. The Moon will pass a bright object, which you can identify as Venus with the Object Identification Tool (the arrow icon in the icon/location box; if this box is not visible on the screen, activate it by clicking Show Floating Palettes in the Window menu). To remove the "Venus" label that appears when this tool is used, click again on blank sky. Advance time in 1-day steps to watch the Moon grow from crescent toward quarter moon as it moves in its orbit.

Change back to the Grabber Tool (click on the hand icon in the icon/location box). It is advisable at this stage to move your viewpoint around the horizon with the Grabber Tool to place the Moon in the center of the screen, since the enlarged moon image is distorted when it moves close to the edge of the screen. You can now advance time again by several days to July 27, using the Grabber Tool to move the screen westward (toward the right) as necessary to keep the Moon in view. You will see the full transformation from crescent through quarter to gibbous and then to full moon.

The full moon will appear in the East at sunset and will just be rising as the Sun is setting. The reason for this is that, in order for us to see a fully illuminated Moon, the Sun must be behind us when we face the Moon; that is, the Sun must be in the opposite part of the sky from the Moon.

It is perhaps instructive to repeat this procedure with the ecliptic illuminated (Guides/Ecliptic or CTRL+3), to follow the Moon's path with respect to the Sun's track across our sky. Again, you will see that the Moon's path does NOT follow precisely that of the Sun, but crosses it twice per orbit at the nodes.

B. Waning Phases

Screen Set-up

1. Change the time to midnight, 12:00:00 AM. Make sure you set AM rather than PM.
2. Advance the date to July 28, 1999 (7/28/1999 AD).
3. Move your viewpoint to S.

Procedure

Again, advance time in 1-day steps and follow the Moon through the waning phases of its cycle, using the Grabber Tool to move along the horizon as necessary to keep the Moon in the center of the screen. You will note that the Moon passes Jupiter on August 4 (8/4/1999) as it approaches the eastern horizon.

To watch the late phases of this cycle, advance the time to dawn, 4:45:00 AM, move your viewpoint to E, and advance time in 1-day steps again. During this phase, the Moon will rise before dawn. You can show that the new moon occurs on about August 11.

Questions

1. How many days elapse between
   a. new moon and first quarter?
   b. new moon and full moon?
   c. new moon and the next new moon?

2. In which way do the "horns" of the crescent moon point in the early phases, with respect to the direction to the Sun?

3. In which way do the "horns" of the crescent moon point in the late lunar cycle phases, with respect to the direction to the Sun?

4. Above which horizon do we see the full moon at its highest in the sky?

PART 3. Setting of the Sun and Crescent Moon

It is perhaps instructive to show a sunset sequence with a new crescent moon nearby. This is one of the finest sights that occurs every month: The thin crescent, with the rest of the Moon faintly illuminated, slowly slides below the horizon in a dark blue twilight.

A. Screen Set-up

1. Set date to July 14 (7/14/1999 AD) and the time to 7:15:00 PM, just before sunset.

2. Move your viewpoint to W.

3. Set the time interval to 1 minute.

4. Remove the trees from along the horizon by clicking on Settings, then on Options. Click on the down arrow to see a list of options, click on Horizon, click off the Scenery, and then click OK to return to the screen.

B. Procedure

Advance time continuously in 1-minute steps and watch the Sun and then the Moon slowly set over the western horizon. Venus can be seen easily and the star Regulus becomes visible as the twilight sky darkens through sunset. Return to 7:15:00 PM and re-run this sequence, stopping the motion and noting the times of setting of the Sun, the Moon, Venus, and Regulus.

Questions

1. What is the time of sunset on this date?

2. What is the time of moonset?

3. When does Venus set?

4. When does Regulus set?

PART 4. Lunar Occulation of a Star

One striking consequence of the lunar motion is the occasional passage of the Moon in front of a star, known as an occultation. This is an exciting event to watch through a telescope, particularly if circumstances are such that only the Moon's northern or southern polar regions pass over the star in what is known as a grazing occultation. In this case, the star is seen to wink out several times as it moves behind the mountains and crater walls of the polar regions of the Moon.

It is possible to simulate an occultation with the present configuration, though not a grazing event. This demonstrates strikingly the motion of the Moon against a star background.

A. Screen Set-up

1. Set your location to the North Pole (Settings/Viewing Location/Others/North Pole/Set Location).

2. Set the date to July 29, 1999 (7/29/1999 AD).

3. Set the time to 8:00:00 AM.

4. Set the time step to 1 minute.

5. Switch off daylight by clicking on Daylight in the Display menu.

6. Set the field of view to 4° using the Magnification Tool, as follows. Click on the magnifying glass icon in the icon/location box or press F8 to activate the Magnification Tool. Then place the magnifying glass near the center of the screen (you may find it interesting to place it on the Pleiades, the little cluster of stars near the center of the screen), and click the left mouse button several times while watching the field of view displayed in the icon/location box. Stop when the field of view is 4°. If you wish, at any time you may demagnify the image by holding the Control (CTRL) key on the keyboard while clicking the Magnification Tool.

7 Lock the field of view onto the Moon. To do this, highlight the Moon in the planets box (if this box is not already on the screen then click Window/Planets to activate it, and click on the triangles next to the Sun and the Earth if necessary to see the Moon in the list), and click on the Center and Lock button. Note that if you adjust the magnification using the Magnification Tool after this, you will need to re-engage this Center and Lock facility. However, if you change the field of view by clicking on the field-of-view number in the icon/location box, the Center and Lock condition will be maintained.

You can now see the Moon at its true size for this field of view, whether the "enlarge moon image" is set or not. You can also see a star to the left or lower left of the Moon.

If we had not locked onto the Moon, then if we were to advance time, even by a small amount, the whole scene would change drastically and the Earth's rotation would carry the Moon completely out of our field of view. Locking onto the Moon is like looking through a telescope that is steering so that the Moon stays centered in our field of view.

Setting our location to the North Pole means that our angle of tilt relative to the Moon remains constant as the Earth rotates. If we had not done this, then the sky (and Moon) would seem to rotate on the screen as time passed.

B. Procedure

Click the time-start button in the date/time box, and watch the occultation. The star is occulted by the southern part of the Moon.

As you watch the occultation, it may look as if the Moon is at rest and the stars are moving past the Moon; but remember that it is really the Moon that is moving past the stars. The illusion of moving stars arises because we have locked our field of view onto a moving Moon.

In real life stars are (almost) true points of light in the sky, and therefore they disappear almost instantly as the Moon's limb covers them. In fact, a star's angular size can sometimes be measured by the careful measurement of the disappearance time of a star during such an occultation! However, the simulation shows a small disk for the star, and a rather slow disappearance. You can reverse time and see the event several times, through the magic of the computer!

One other important measurement can be made using a grazing occultation of a star by the Moon. If several observers, spaced out at right angles to the shadow path of the Moon across the Earth, watch the occultation through telescopes and make careful timings of the disappearances and reappearances of the star, then these timings can be used to construct a detailed profile of the mountains along the Moon's limb.

A similar procedure can be used when an asteroid occults a star. Asteroids are too far away for us to be able to see them as much more than just points of light, even through the largest telescopes. However, occultation observers scattered across the shadow path of the asteroid on Earth see different parts of the asteroid passing in front of the star, so careful timings of the disappearance and reappearance of the star can tell us the shape and size of the asteroid. Occasionally, we see the star disappear a second time and then reappear, telling us that the asteroid has a companion moon orbiting it.

Answers

Moon's Motion

1. Toward the east.

2. At about 13° per night.

3. Curved, close to but not exactly along the ecliptic.

Phases

1. a. about 7 days; b. about 14 days; c. about 28 days.

2. The horns point away from the Sun.

3. The horns point away from the Sun.

4. The southern horizon.

Setting of the Sun and Moon

1. Sunset: 7:27 PM (standard time).

2. Moonset: 8:36 PM (standard time).

3. Venus sets: 9:12 PM (standard time).

4. Regulus sets: 9:15 PM (standard time).

5

The Earth's Orbital Motion and the Sun's Apparent Motion Across the Sky

Our view of the sky is constantly changing. The two most evident changes are the apparent motion of the stars and constellations toward the west over the course of a single night, and the much slower shift of stars and constellations toward the west over the course of a year when viewed at the same time each night (say, 11 PM). The first of these changes is caused by the Earth's rotation, which makes the sky appear to rotate around the Earth's polar axis once every day, or 15° per hour; it also makes the Sun, Moon, and stars rise in the east and set in the west each day. The second change is caused by the Earth's orbital motion around the Sun; since the Earth makes a full orbit of 360° in about 365 1/4 days, our view of the universe at any given time of day or night changes by about 1° per day. The combined result of these two changes is that each star rises about 4 minutes earlier each night.

If we could see the stars in the daytime, then the most obvious effect of our orbital motion around the Sun would be an apparent motion of the Sun past the background stars. Unfortunately, this motion is hidden from us by the blue sky of daylight (the stars are not visible when the Sun is up). By the magic of computers, however, we can simply turn off the blue sky to reveal the Sun's apparent motion against the background sky!

The plane containing the Sun and the Earth's orbit is known as the **ecliptic plane**, and the projection of this plane onto the sky forms a line or circle around the sky called the **ecliptic**. This path passes through a limited set of constellations known as the **Zodiac**, a band of constellations through which most of the brightest planets and the Moon appear to move as they orbit close to the ecliptic plane. This region of our sky played an important role in early astrology because the positions of the planets were used to predict the future, and these zodiacal constellations or "signs" are still used in modern astrology.

Some more information on the Earth's orbit and the apparent motion of the Sun can be found in the discussion of the seasons in Section 2-5 and Figure 2-13 of Kaufmann and Freedman, *Universe*, Fifth Ed.

This exercise will demonstrate the daily motion of the Sun past the background stars, and show the Sun's path—the ecliptic—around the sky.

A. Screen Set-up

1. After starting the program, check the icon/location box to see that the field of view is 100° (if this box is not visible on the screen, you can activate it by clicking Show Floating Palettes in the Window menu). If the field of view is not 100°, then click on the house button to reset the field of view to 100°.

2. Set the location to Chicago (Settings/Viewing Location/North America/United States/New York/Set Location).

3. If your viewing direction is not already toward the south (S), then click the S button on the button bar near the top of the screen. If this button bar is not visible, activate it by clicking Toolbar in the Window menu. After the screen adjusts direction, you should see the symbol S on the horizon near the bottom of your screen.

4. Check that the Daylight Saving Time option is turned off. To do this, check the small sun icon to the left of the time in the

date/time box (if this box is not visible on the screen, activate it by clicking Time in the Window menu). The sun icon is yellow with 8 rays if DST is activated, but background color with 4 rays if DST is not activated. If the icon is yellow, then click on it to deactivate DST.

5. Stop the time by clicking the stop-time button in the date/time box; then
   a. set the date to December 21, 1999 (12/21/1999 AD);
   b. set the time to midday, 12:00:00 PM;
   c. set the time interval in the date/time box to 1 sidereal day. Be sure that the time units are sidereal days, not just days (which means solar days).

   The reason for setting the interval to 1 sidereal day is that the background sky returns to exactly the same position after 1 sidereal day, since 1 sidereal day is the true Earth rotation period with respect to the background stars. The stars then appear to be motionless as we advance time, making the motion of the Sun easier to see.

6. Switch off daylight by clicking on Daylight in the Display menu. You can now see the background stars behind the Sun.

B. The Sun's Motion

1. Step time forward by one sidereal day at a time with the single-step button in the date/time box. You will notice as you do this that the time displayed in the date/time box, which is Solar Time, actually advances by less than 1 day since 1 sidereal day is less than 1 solar day by 3 minutes, 56 seconds. (I.e., 1 sidereal day is equal to 23 hours, 56 minutes, 4 seconds in Solar Time.)

2. Set time running forward by clicking on continuous run (the button to the right of the stop button in the date/time box). Experiment with time, moving it forward and backward to see the Sun's apparent motion.

3. Although the background stars remain fixed as you advance time by 1 sidereal day steps, you may notice several star-like

objects that move across the sky at a different rate than the Sun. These are, of course, the **planets**, orbiting the Sun at their own individual rates. You can identify each one by using the Object Identification Tool (the arrow in the icon/location box; if this box is not visible on the screen, you can activate it by clicking Show Floating Palettes in the Window menu) and by using the arrow icon to click on and label each object in turn. (To remove the final label, click on a blank patch of sky.)

4. Display the orbit of each planet as follows. Click Planets in the Window menu to display the planet box, and look to see if there is a list of planets; if not, then click on the triangle next to the Sun to display the planet names. For each of your identified planets in turn, click on the Orbit Column next to the planet's name to display the orbit, then click here again to remove the orbit from the screen. **Mercury** and **Venus** are close to the Sun and move in orbits that are more or less inclined to the ecliptic plane and hence do not follow the ecliptic exactly. The **Moon** also makes its appearance as you advance time, moving much more quickly across the sky every sidereal day than does the Sun. You can display the Moon's orbit by clicking on the Orbit Column next to the Moon's name; if the Moon is not visible in the list of planets, click on the triangle next to the Earth to display the Moon's name. When you are finished, you can click on the X at the top of the box to remove the box from the screen.

5. Run time backward until the Sun is against the right edge of the screen. Then step the time forward and count the number of sidereal days that elapse until the Sun reaches the left edge of the screen, having traveled about 100° across the sky. Divide 100° by the number of days taken for the Sun to move through this angle to obtain its motion per day. (For this rough estimate, you can assume that 1 sidereal day is equal to 1 solar day.)

6. The Sun's motion appears smooth, with evenly spaced daily steps, but in fact, its apparent speed across our sky varies because the Earth moves in an elliptical orbit, moving faster at some times and slower at others. This variation is discussed and illustrated in the exercise on the analemma.

7. In the steps above, you advanced the time by one sidereal day per time-step to keep the stars fixed in the sky. This allowed

you to see the apparent motion of the Sun relative to the stars. You may find the following exercise interesting and fun to try.

a. Starting from the position at the end of the previous step, run time backward until the Sun is at or close to the center of the screen, then click the stop button.
b. Change the time step from sidereal days to days (i.e., solar days).
c. Click the continuous run button. Because of the 1-solar-day time interval, you should see the Sun remain almost fixed in position while the background stars move past it. (Changes in the Sun's position at noon from day to day are discussed in the exercise on the analemma.)

As you watch this, imagine yourself on the outside edge of a merry-go-round, watching the Sun at the center of the merry-go-round. Since you are watching the center, the center stays fixed in your view while the rest of the universe continuously revolves around you. For the same reason, the constellations that we see at night shift slowly toward the west in a one-year cycle when we watch the sky from day to day (or night to night)(being on the Earth as it orbits the Sun is like being on a gigantic merry-go-round that takes one full year to go around once.

Questions

1. In which direction did the Sun appear to move in the New York sky from day to day, relative to the background stars, as the Earth orbited the Sun? How does this direction compare to the apparent direction of motion of the Sun from hour to hour in a single day, due to the Earth's rotation?

2. What was the shape of the path taken by the Sun (e.g., straight, curved)? This path defines the ecliptic plane. Switch on the ecliptic by clicking Ecliptic in the Guides menu, or press CTRL+3, to see this path displayed for you by the program. At this stage, you can use the Grabber Tool (with the hand cursor) to move around the horizon to see the shape of the ecliptic over other portions of the sky.

3. Approximately what was the solar motion in degrees per day across the background sky?

4. When you used solar day time intervals, in which direction did the background stars appear to move from day to day relative to the Sun (toward the east or toward the west)? Toward which direction, therefore, does the Earth move in its orbit around the Sun, as seen by someone facing the Sun at noon? (It may be easier to answer this question if you again imagine yourself at the edge of a merry-go-round and think about what you would see happening to the background scene if you were to keep your eyes on the center.)

Answers

1. Toward the east. Opposite to the daily motion, since the Sun appears to move toward the west from sunrise to sunset each day.

2. Curved.

3. Approximately 1° per day.

4. Toward the west.

6

The Sun's Position at Midday and the Analemma

The Sun's position at midday at any site will not always be on the observer's meridian. One reason for this is that the site may be offset from the center of the time zone (i.e., the site is not on the central meridian of the time zone); this produces an offset of the Sun from the observer's meridian at noon that remains constant throughout the year. If time zone boundaries were ideal, then this offset could be as large as 1/2 hour; but in fact time zone boundaries sometimes deviate from ideal for political reasons, resulting in offsets of up to about an hour for some locations.

In addition to this constant offset, there is also a varying deviation of the Sun from the meridian that is caused by the combination of two effects. The first of these is the Earth's elliptical orbit, which results in a variable speed of the Earth around the Sun, as described by Kepler's second law. This law is described in Section 4-4 of Kaufman and Freedman, *Universe*, Fifth Ed. The changing orbital speed of the Earth around the Sun changes the apparent speed of the Sun across our sky, day by day. The second effect is the tilt of the Earth's spin axis to its orbital plane, also called the ecliptic plane. This also causes a variation in the speed of the apparent motion of the Sun.

The varying deviation of the Sun east or west described in the previous paragraph, when combined with the annual motion of the Sun north and south due to the seasons, results in a figure-eight pattern for the Sun's midday position in our sky over the course of a year. This figure-eight pattern is called the **analemma**, and is often found on globes representing the Earth. This truly amazing multiple-exposure photograph required a year to take, at a rate of one exposure approximately every 8 days!

The shape of the analemma became important when sundials were used to tell the time by tracing the movement of the shadow of a vertical post upon the ground. The analemma provided corrections, amounting at times to 16 minutes, to times determined by this shadow at different times of the year.

The present exercise demonstrates the analemma by tracing the apparent motion of the Sun in the sky at midday for a whole year at a chosen site.

PART 1. Finding the Shape and Size of the Analemma

A. Screen Set-up

1. After starting the program, check the icon/location box to see that the field of view is 100° (if this box is not visible on the screen, you can activate it by clicking Show Floating Palettes in the Window menu). If the field of view is not 100°, then click on the house button to reset the field of view to 100°.

2. Set the location to Calgary, Alberta, Canada, near 50° latitude (Settings/Viewing Location/North America/Canada/Calgary/Set Location). At this latitude the Sun is always visible on the screen at noon; that is, it is above the horizon in midwinter and not out of view above the top of the screen in midsummer.

3. If your viewing direction is not already toward the south (S), then click the S button on the button bar near the top of the screen. If this button bar is not visible, activate it by clicking Toolbar in the Window menu. After the screen adjusts direction, you should see the symbol S on the horizon near the bottom of your screen.

4. Check that the Daylight Saving Time option is turned off. To do this, check the small sun icon to the left of the time in the date/time box (if this box is not visible on the screen, activate it by clicking Time in the Window menu). The sun icon is yellow with 8 rays if DST is activated and background color with 4 rays if DST is not activated. If the icon is yellow, click on it to deactivate DST.

5. Stop the time by clicking the time-stop button in the date/time box; then
   a. set the date to the winter solstice, December 21, 2000 AD (12/21/2000 AD);
   b. set the time to midday, 12:00:00 PM;
   c. set the time interval in the date/time box to 7 days. (Make sure you use "days," which actually means solar days and not "sidereal days," since you want to trace the Sun's apparent motion at midday every 7 solar days.)

6. Switch off daylight by clicking on Daylight in the Display menu. This will enable you to see the stars moving across the daytime sky, week-by-week.

7. Check that the cursor is the Grabber Tool (hand icon). If it is not, then select the Grabber Tool by clicking on the hand icon in the icon/location box. If this box is not visible on the screen, you can activate it by clicking Show Floating Palettes in the Window menu. Alternatively, you can activate the Grabber Tool by pressing the F4 key.

B. Tracing the Analemma

After the above set-up, the Sun should be at a low angle above the southern horizon. You will notice that the Sun is *not* due south at this time. This is because Calgary is offset from the center of its time zone, the Mountain Standard Time zone, by about 36 minutes. You can adjust for this offset, if you desire, by changing the local time to 12:36:00 PM.

The shadows of the trees are cast across the ground before you. You can think of these trees as representing a sundial for measuring time at this location. The position of the Sun in the sky will move these shadows around, as we shall see.

You need to move your viewpoint so that the Sun is just above the bottom of the screen at this time so that you can follow the path of the Sun through the full year. To do this, use the Grabber Tool to move the sky downward (click and hold the left mouse button while sliding the mouse toward you).

In order to trace the motion of the Sun week by week, you need to tape a piece of transparent plastic or tracing paper to your screen. Mark the corners of your screen on the transparent sheet with a felt pen or marker. Be very careful not to damage the screen when you do this. Mark the Sun's position, then single-step the time forward by 7 days and mark the Sun's position again. Continue this process for a whole year of time, tracing out the Sun's position, or the analemma. Leave the plastic or tracing paper where it is for the next step.

Switch on the equatorial coordinate grid (click Equatorial Grid in the Guides menu or press CTRL+2), and mark the Right Ascension grid lines on either side of the analemma and all of the declination grid lines in the vertical direction. These are 1-hour right ascension and 10° declination intervals. Now remove your plastic film or tracing paper from the screen so that you can make measurements on it.

Questions

1. The reason for the north–south motion of the Sun is the tilt of the Earth's axis to the ecliptic plane. Using the declination lines on your plot, estimate this tilt by measuring the maximum N–S excursion of the Sun in degrees of declination over a full year. This will be twice the tilt-angle of the Earth's axis. It is this change in the angle of the Sun in the sky that produces seasonal changes upon the Earth.

2. Estimate the maximum error in a sundial from the width of your plot, assuming that the Sun would be "on time" if it were on the center line that runs lengthwise through the analemma. This makes the sundial error equal to one half of the width of your analemma. For reference, one interval of Right Ascension on the screen is equal to 1 hour of time, or 60 minutes, since one hour of Right Ascension is the angle through which the Earth rotates in 1 hour.

3. At what times of the year will the sundial be most accurate?

4. At what times of the year will the sundial be most inaccurate?

PART 2. The Analemma from Other Latitudes

The shape of the analemma will be the same from any position on Earth. You can test this hypothesis by moving to another location, say Sao Paulo, Brazil (Settings/Viewing Location/South America/Brazil/Sao Paulo/Set Location). Reset the date to 12/21/2000 AD, and the time to 12:00:00 PM. Check that daylight savings time is off. Remember that the Sun will now be in the northern sky from the southern hemisphere so click on the N position on the top toolbar. In order to see the Sun, you will have to use the Grabber Tool to move the sky down so that the Sun is just visible at the top of the screen at this time. Now set time running at 7-day intervals to see the analemma for Sao Paolo. As you see, it is the same shape as it was from a northern latitude site.

PART 3. Sunrise Time at the Winter Solstice

A question that is sometimes asked is, "If the Sun is at its furthest South at the time of solstice, on the shortest day of the year, December 21, why is this *not* the date of latest sunrise?" In fact, the date of latest sunrise is close to the last day of the year, several days later!

In the steps below, you can answer this question in terms of the analemma and the position of the Sun as it moves through the time of the solstice.

A. Screen Set-up

1. Set up again for Calgary, Canada (Settings/Viewing Location/North America/Canada/Calgary/ Set Location).

2. Reset the date to 12/21/2000 AD.

3. Set the time to 8:42:00 AM.

4. Set the time interval to 1 solar day (the time-step labeled "days").

5. Check that Daylight Saving Time is off.

6. Check that daylight is off.

7. Move to the E position using the top toolbar. You should see the Sun on the horizon near the SE position.

8. In order to demonstrate this sunrise effect, it is best to magnify the view of the sky in the region of the Sun. To do this, use the Magnification Tool (activate the Magnification Tool by clicking on the magnifying glass icon in the icon/location box, or press F8). Place the Magnification Tool on the Sun and click several times until the Field of View is 4°, as shown by the number in the lowest box in the icon/location box. Be sure the Magnification Tool is on the Sun each time. If you lose the Sun from the screen at any magnification, then reverse the magnification by holding down the control key (CTRL) while you click the mouse, until you find the Sun again; then re-magnify by clicking on the Sun. When you are finished, return to the Grabber Tool (click on the hand icon in the icon/location box or press F4).

9. You should see the Sun near the center of the screen, just touching the horizon. If necessary, use the Grabber Tool to center the Sun on the screen. If the lower edge of the Sun is well below the horizon or above the horizon, temporarily set the time interval to 10 seconds and step forward or backward to place the Sun at a position just touching the horizon; then reset the time interval to 1 day.

B. Date of Latest Sunrise

1. You can see that on the winter solstice the Sun is just rising at Calgary at the time shown on the screen. Now, advance the time by 1-day steps. You will see that, for several days after December 21, the analemma motion of the Sun places it further below the horizon at 8:42:00 AM than it was the previous day. This means that sunrise will be later on these days than on the shortest day.

2. Experiment to find the date of latest sunrise, and record this date.

3. Set the date to the latest sunrise date, then change the time interval to 10 seconds and step time until the Sun is just clear of the horizon. Record this time.

Questions

1. On what date does the latest sunrise occur?

2. About how many days after the winter solstice does the latest sunrise occur?

3. By how many minutes in the day is this sunrise later than sunrise at the solstice?

Answers

Tracing the Analemma

1. Tilt of Earth's axis = 23 1/2°

2. About 14 minutes.

3. At the solstices (mid-summer and mid-winter), and around May 20 and October 26.

4. About March 11 and October 10.

Date of Latest Sunrise

1. About December 29.

2. About 8 days after the solstice.

3. 2 minutes, 20 seconds.